HARCOURT

Math

GEORGIA EDITION

D0933174

Practice/ Homework Workbook

Grade K

Harcourt

Visit *The Learning Site!*
www.harcourtschool.com

ISBN 13: 978-0-15-349538-0
ISBN 10: 0-15-349538-3

9 10   1421   15 14 13 12 11 10

CONTENTS

Name ____________________

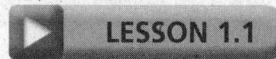

Above, Below, Beside

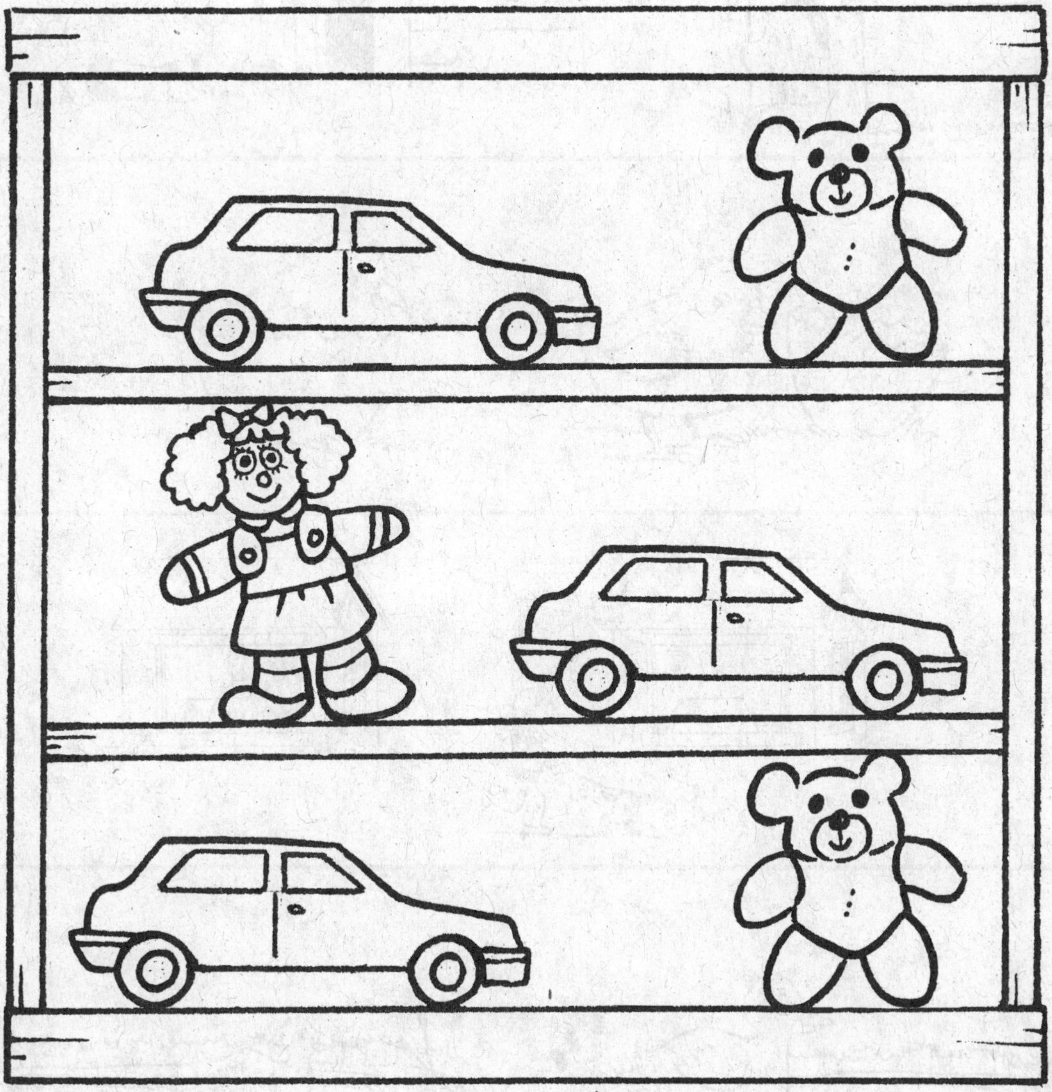

Color the car above the doll. Color the bear below the car.
Color the car beside the doll.

Name _____

Inside, Outside

★

 Circle the dog that is inside the doghouse.

 Circle the chick that is outside the egg.

★ Circle the girl who is inside the house.

♥ Circle the duck that is outside the pond.

PW2   Practice/Homework

Name _____

In Front Of, Behind

 Circle the crayons in front of the books. Mark an X on the pencils behind the books.

 Circle the flowers in front of the girl. Mark an X on the flowers behind the girl.

★ Circle the cat behind the tree. Mark an X on the cat in front of the table.

♥ Circle the object in front of the car. Mark an X on the object behind the car.

Name ____________________

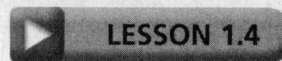

Problem Solving Skill: Use a Picture

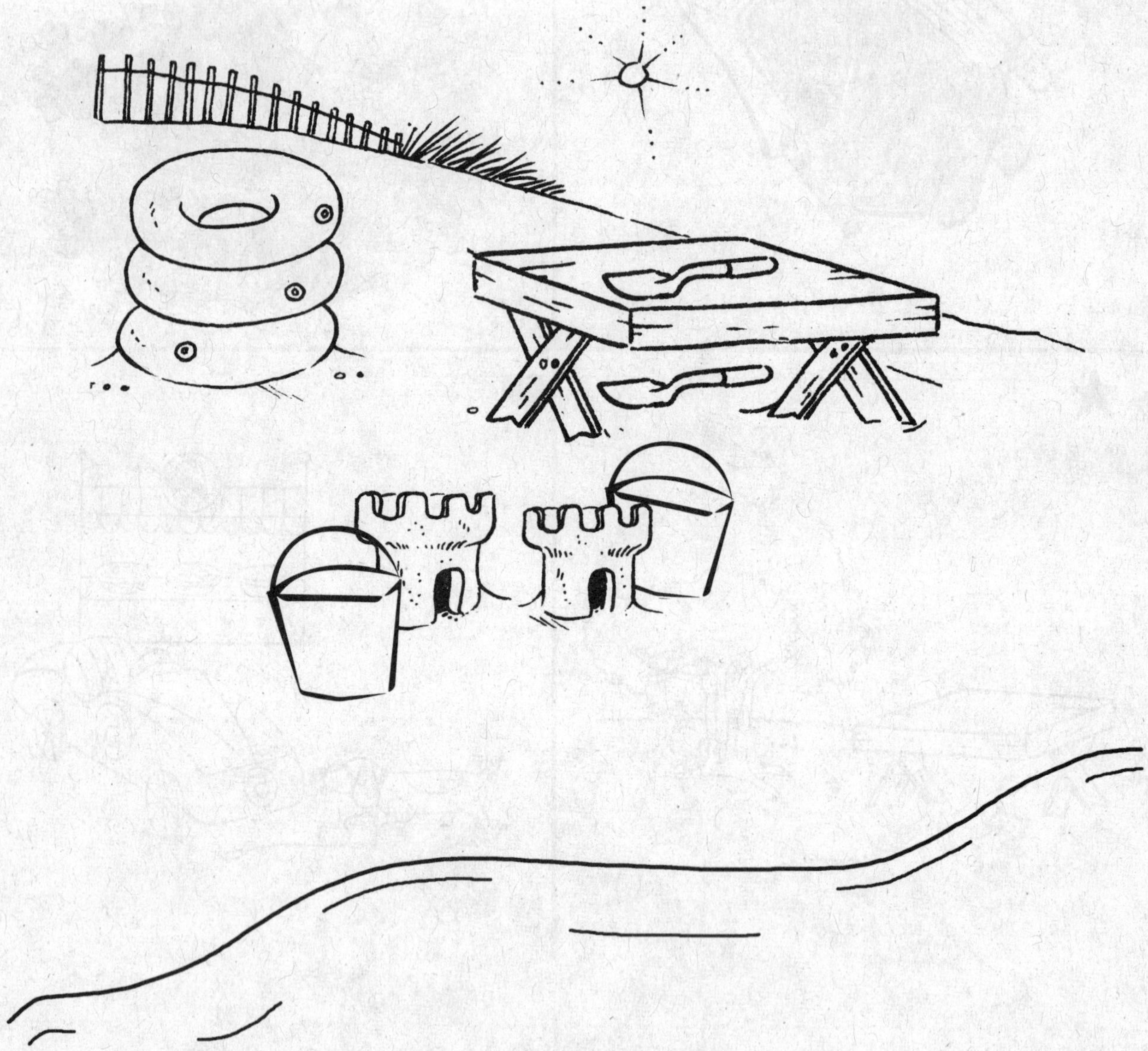

Draw a cloud above the sun. Color the shovel below the table.
Color the bucket in front of the sandcastle. Draw a boy beside the tubes.

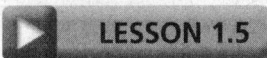

Algebra: Sort by Color or Shape

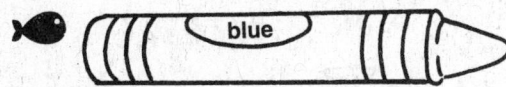

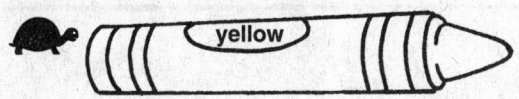

★ □ Square

♥ ○ Circle

Sort small Attribute Links by color.
 Color the crayon blue. Trace and color the blue shapes.
🐢 Color the crayon yellow. Trace and color the yellow shapes.
Sort small Attribute Links by shape.
★ Trace and color the square shapes.
♥ Trace and color the circle shapes.

Name _____

Algebra: Sort by Size or Kind

    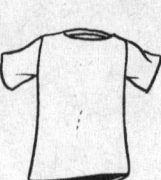

   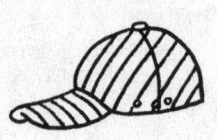  

Circle the objects that are the same size.
Circle the objects that are the same kind.

Problem Solving Strategy: Use Pictures

 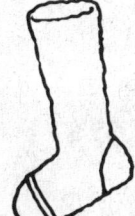  

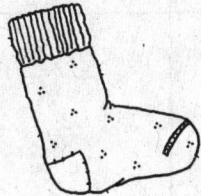 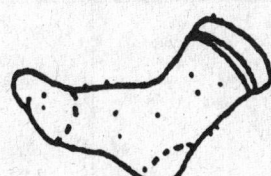 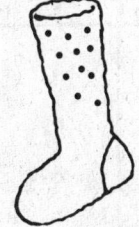 

  ★   Circle the items that are the same kind.

Make a Concrete Graph

Red and Yellow Counters

- Place a handful of counters in the work area. Sort by color.
- Color the counter in the top row of the graph red and the counter in the bottom row of the graph yellow. Move the counters to the graph. What does this graph show?

© Harcourt

Algebra: Movement Patterns

★ ★ ★ Act out the pattern. Say the pattern as you act out each part. Circle what you would most likely do next.

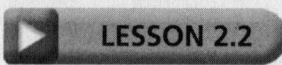

Algebra: Read and Copy Simple Patterns

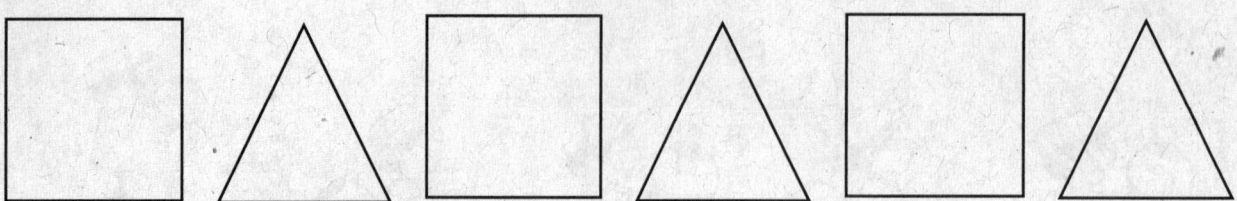

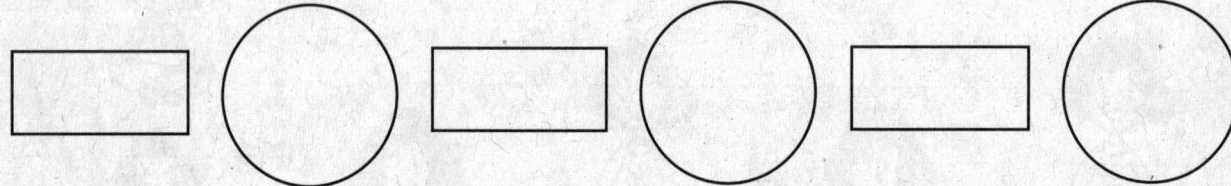

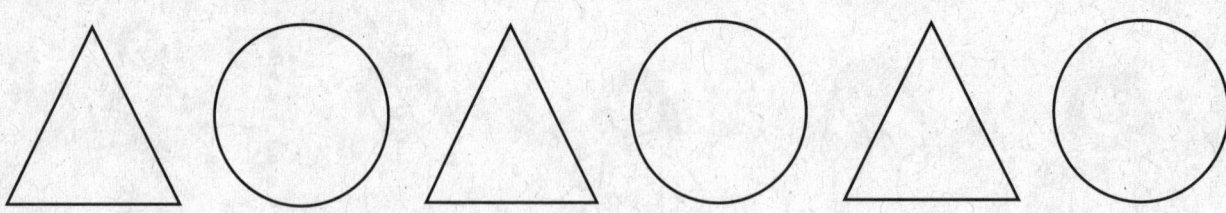

Read the pattern. Draw to copy the pattern.

Name _____

Algebra: Copy and Extend Patterns

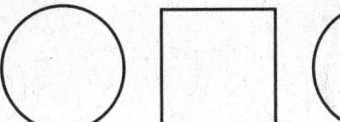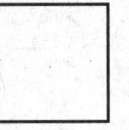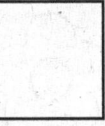

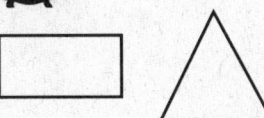

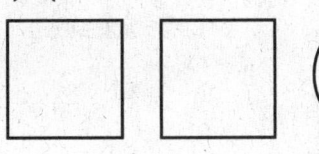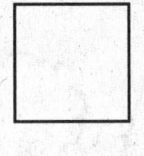

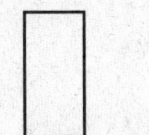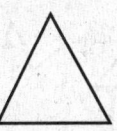

  ★ ♥ Draw the shapes to copy the pattern and show what most likely comes next.

Name _____

Algebra: Predict and Extend Patterns

_____ _____

_____ _____

_____ _____

_____ _____

What pictures do you think come next? Draw the pictures to show what most likely comes next.

© Harcourt

Name _____

Problem Solving Skill: Transfer a Pattern

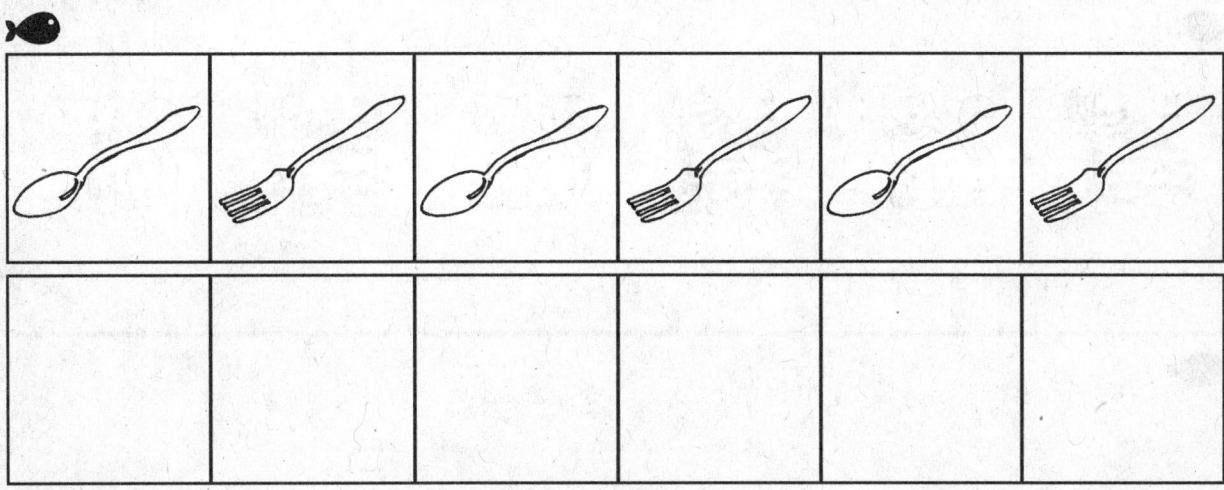

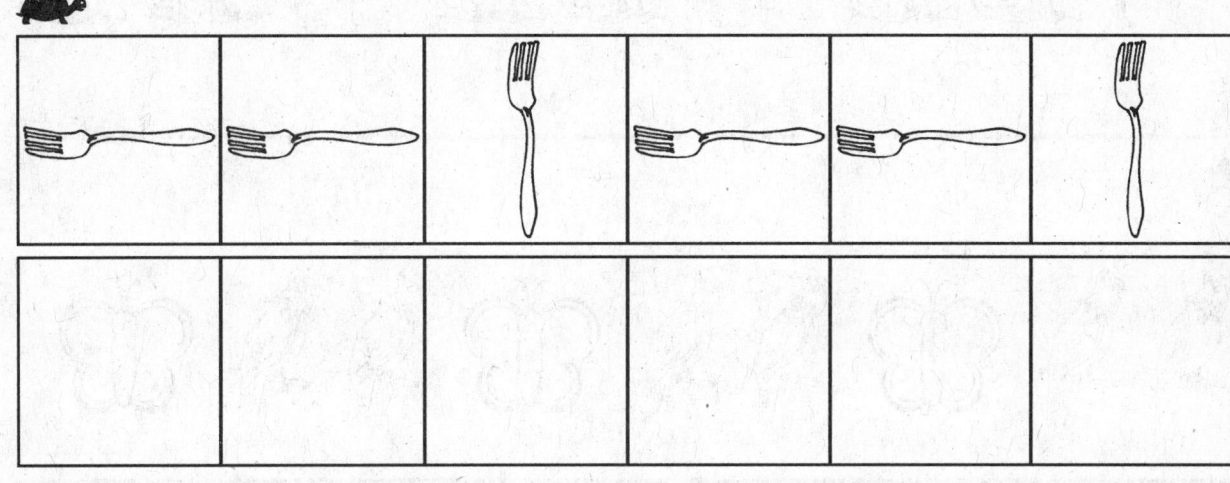

🐟 🐢 ★ Use counters or act out to show the same pattern. Draw the pattern.

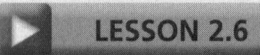
Algebra: Understand a Pattern

 Read the pattern. Circle the part that repeats again and again.

Create a Pattern

🐟

🐢

⬛ Use connecting cubes to make your own pattern. Draw your pattern.

🐢 Use Attribute Links to make your own pattern. Draw your pattern.

Problem Solving Skill: Use a Pattern

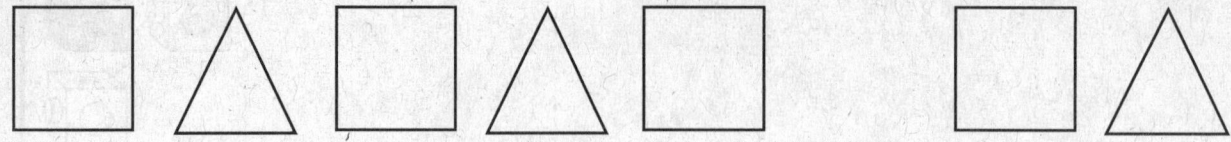

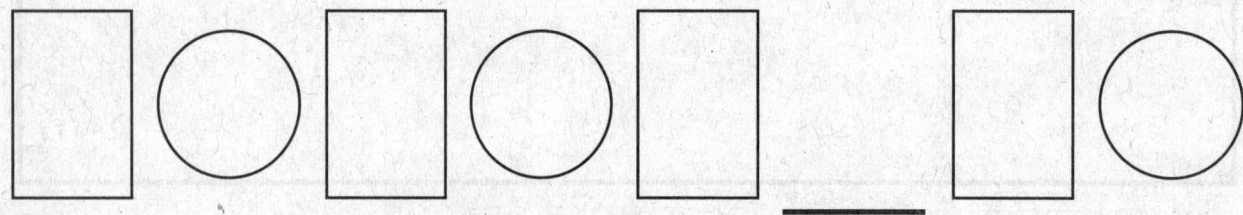

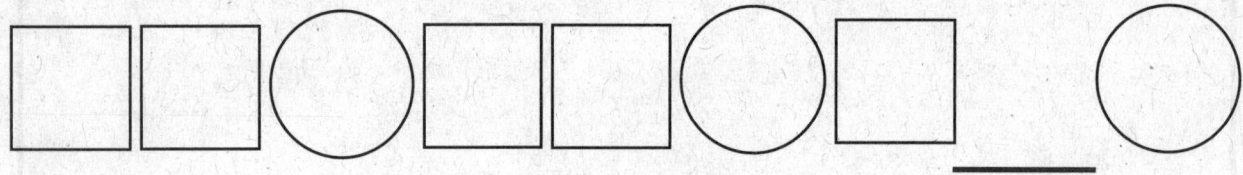

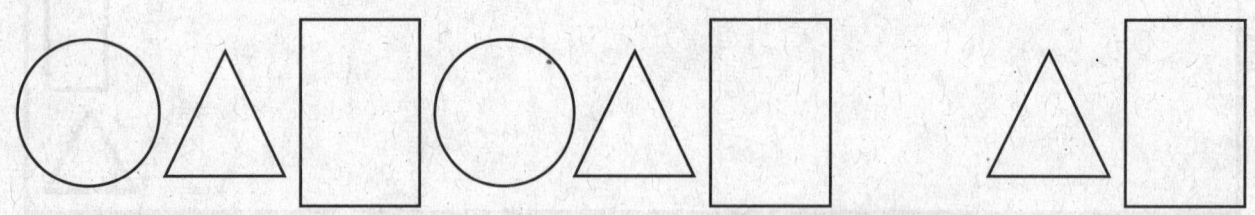

Read the pattern. Tell what part repeats again and again. Draw the missing shape in the pattern.

Name _____

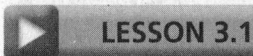

Algebra: Equal Sets

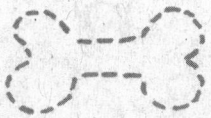

 Draw a bone for each dog.

 Draw a pot for each plant.

© Harcourt

Name _____

Algebra: More

Draw lines to match the animals in the two sets. Compare the sets. Circle the set that has more.

Algebra: Fewer

★

 ★ Draw lines to match the animals in the two sets. Compare the sets.
Circle the set that has fewer.

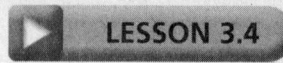

Problem Solving Strategy: Use Objects

Which Group Has More?

Yellow	Red

Put a handful of counters in the bowl. Move the counters to the graph. Draw and color.
Circle the column with more counters.

Name _____

One, Two, Three, Four

1 apple

2 oranges

3 bananas

4 strawberries

Read the number. Draw that many pieces of fruit in the basket.

Name _____

Five

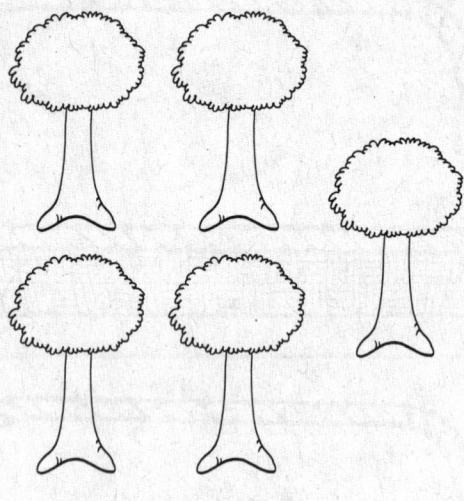

_____

- - - - - - - - - - - -

_____

_____

- - - - - - - - - - - -

_____

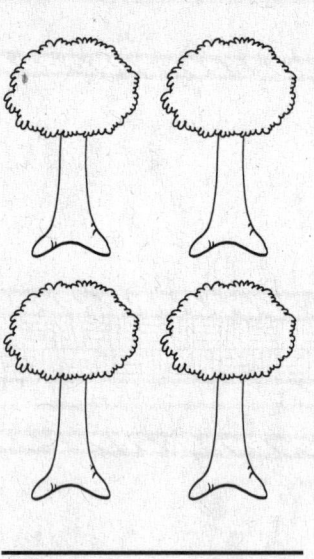

_____

- - - - - - - - - - - -

_____

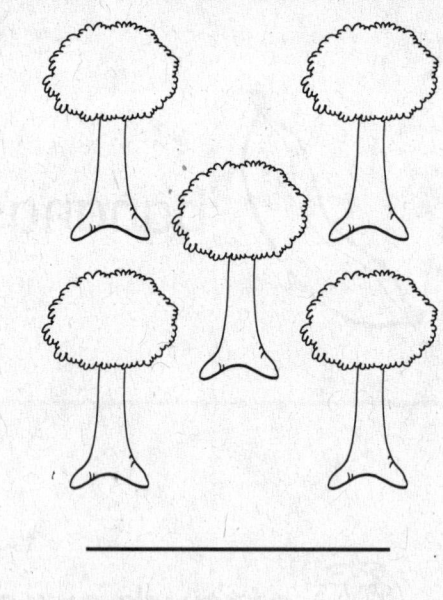

_____

- - - - - - - - - - - -

_____

Count the trees. Write the number that tells how many.

PW 22   Practice/Homework

Name _____

Zero

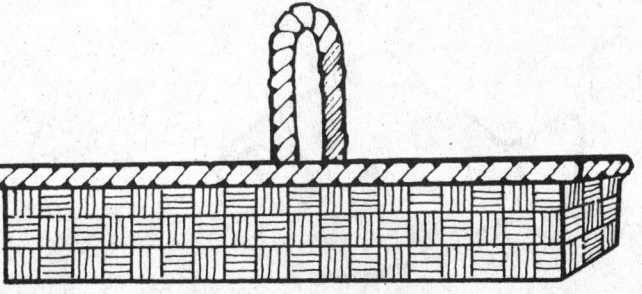

0
zero

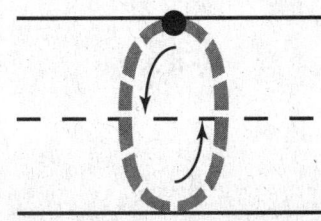

3 4 ⑤

★

0 1 2

♥

1 2 3

❀

0 1 2

🐟 Read the number and look at the picture. Trace the number.

🐢 ★ ♥ ❀ Circle the number that tells how many flowers are in each basket.

Name _____

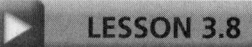

Problem Solving Skill: Use Estimation

Look at the fish at the top of the page. Without counting, use red to circle the fish that have more than five spots. Use blue to circle the fish that have fewer than five spots.

Six and Seven

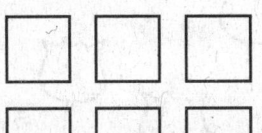

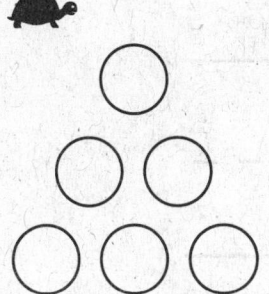
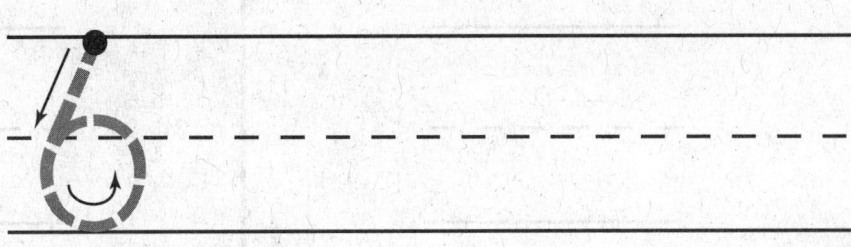

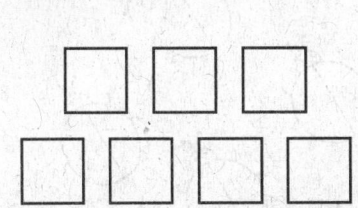
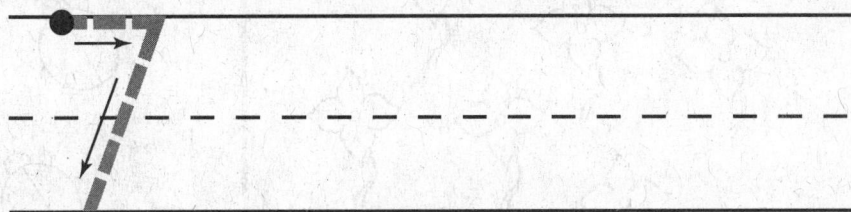

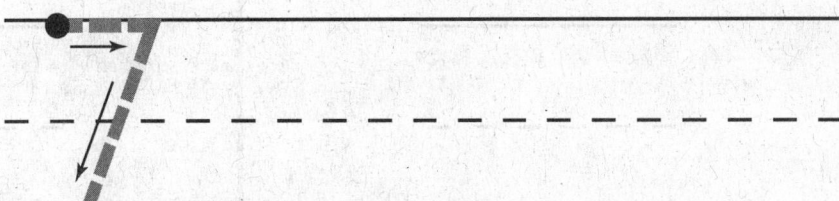

 Count the shapes in the group. Write the number.

Eight and Nine

🐟

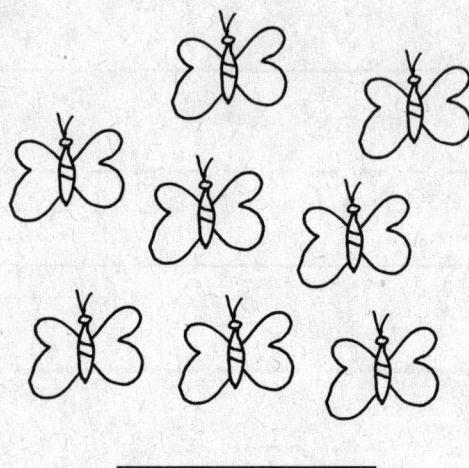

_____

– – – – – – –

_____

🐢

_____

– – – – – – –

_____

★

_____

– – – – – – –

_____

♥

_____

– – – – – – –

_____

🐟 🐢 ★ ♥   Count. Write the number.

© Harcourt

Ten

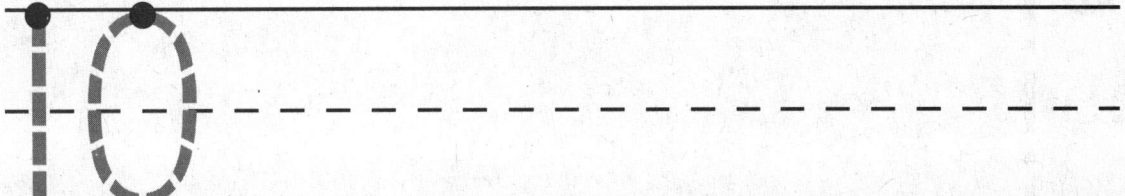

8   9   10

8   9   10

8   9   10

🐟 Write the number 10.

🐢 ★ ❤ Count the bees. Circle the number that tells how many. Write the number.

Practice/Homework   PW27

Name _____

Problem Solving Strategy: Use Objects

Connect ten cubes. Then make one break to get two groups—one with more cubes than the other. Put the groups on the work space. For each group, draw the cubes and write the number. Circle the number that is greater. Use different numbers to make ten each time.

Name _____

Algebra: Share Objects Equally

_____   _____

- - - - -   - - - - -

_____   _____

_____   _____   _____

- - - - -   - - - - -   - - - - -

_____   _____   _____

_____   _____

- - - - -   - - - - -

_____   _____

  ★   Share the objects equally. Draw the same number of objects in each bucket. Write how many are in each bucket.

Name _____

Write Numbers 0 to 10

 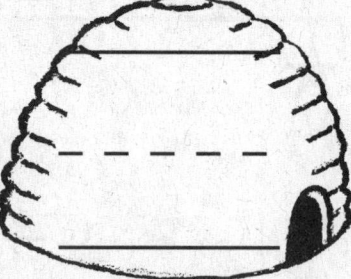

🐟 🐢 ★ ♥ ✿ Count the bees and write the number.

© Harcourt

Algebra: Number Combinations

🐟

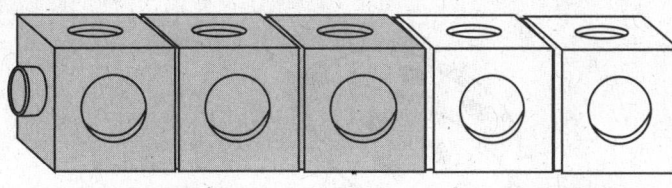

blue — — 3 — —   — — 2 — — yellow

🐢

_____   _____

blue — — — — — —   — — — — — — yellow

★

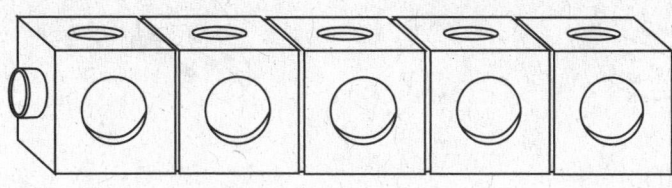

_____   _____

blue — — — — — —   — — — — — — yellow

🐟 🐢 ★   Use blue and yellow cubes to show different ways to make 5. Color the cubes. Write how many of each color.

Problem Solving Skill: Use Estimation

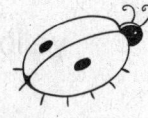

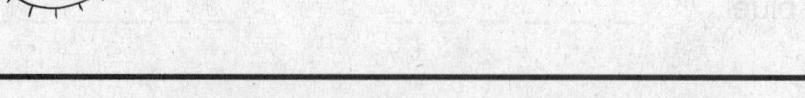 Look at the spots on the big ladybug. It has 6 spots in all. Without counting, find each ladybug that has about 6 spots. Circle the ladybugs that have about 6 spots.

Look at the spots on the big ladybug. It has 8 spots in all. Without counting, find each ladybug that has less than 8 spots. Circle the ladybugs that have less than 8 spots.

© Harcourt

Algebra: Sort Solid Figures

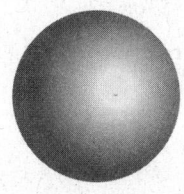   

      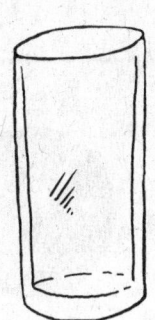

      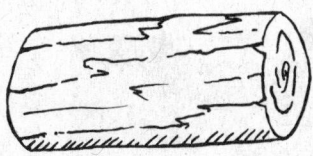

Use blue to circle the objects shaped like spheres.
Use yellow to circle the objects shaped like cubes.

Algebra: Move Solid Figures

 Circle the shapes that slide.
 Circle the shape that stacks.
★ Circle the shapes that roll.
♥ Circle the shapes that roll and slide.

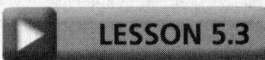

Problem Solving Skill • Use Visual Thinking

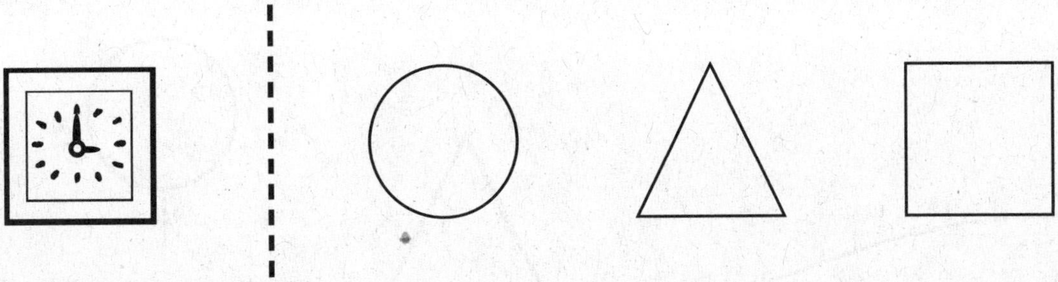

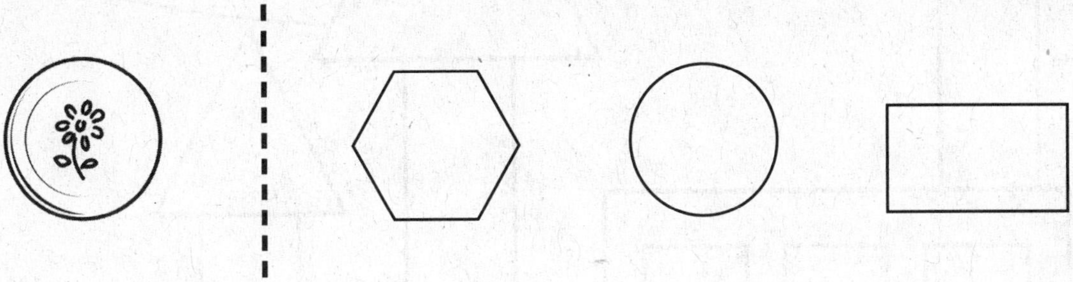

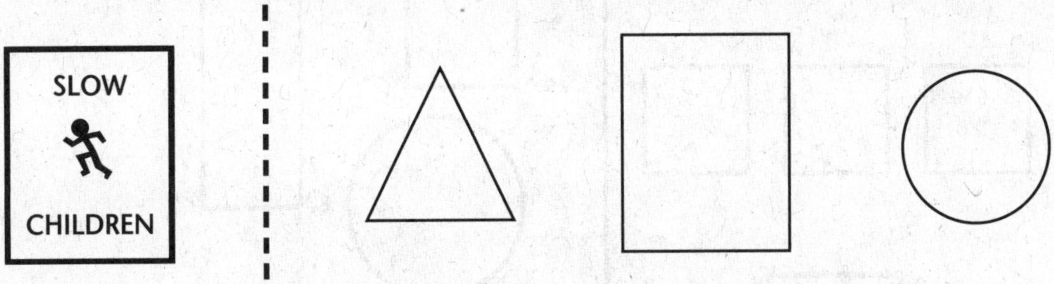

🐟

🐢

★

SLOW

CHILDREN

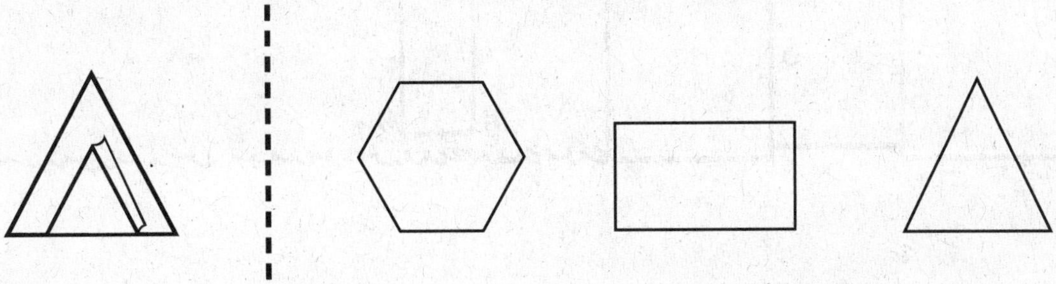

❤

🐟 🐢 ★ ❤ Look at the object at the beginning of the row. Circle the outline that matches the shape of the object.

Name _____

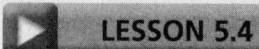

Algebra: Sort Plane Shapes

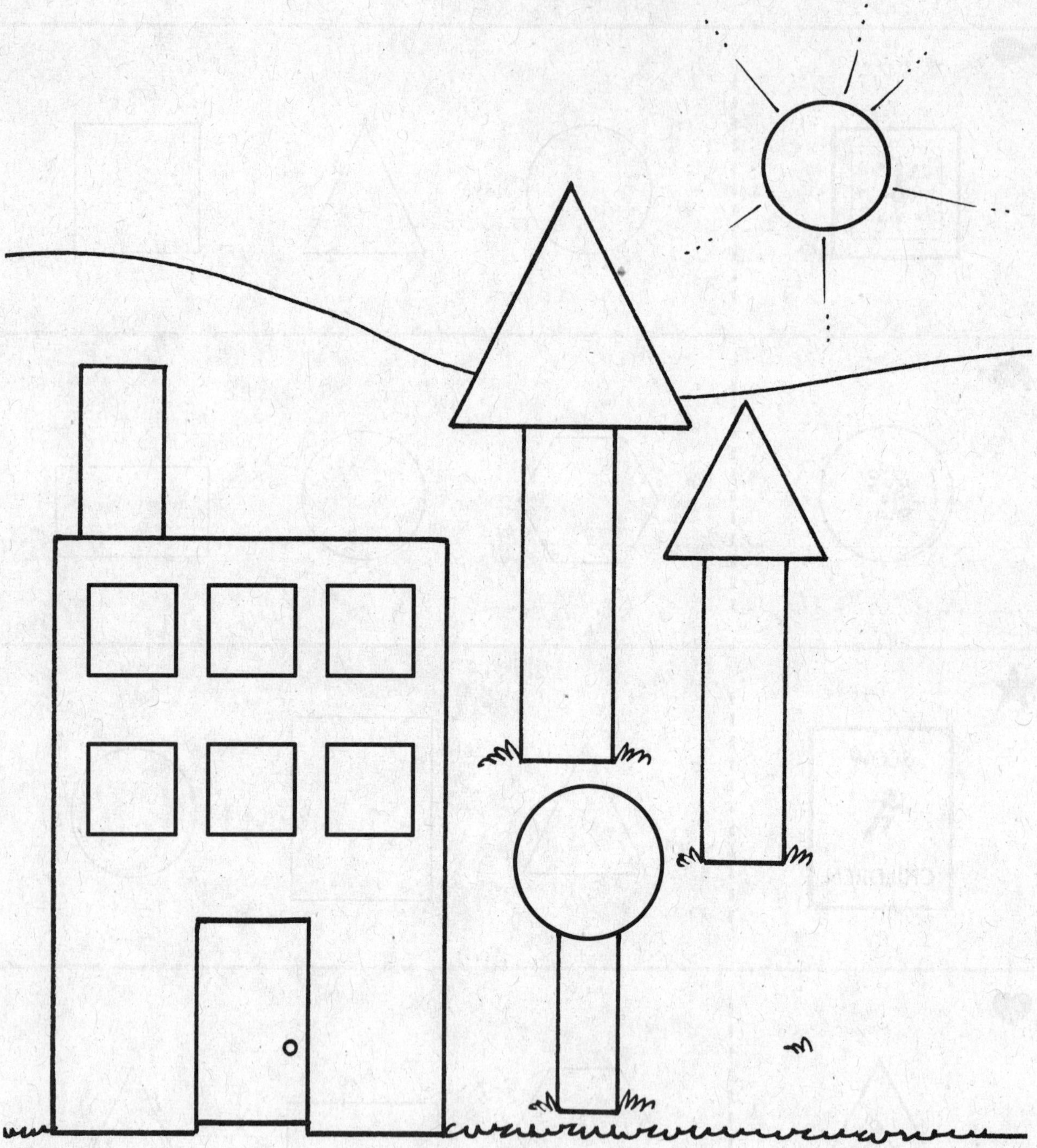

Color the circles yellow. Color the squares blue. Color the rectangles brown.
Color the triangles green.

© Harcourt

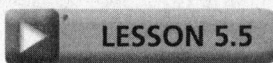

Plane Shapes in Different Positions

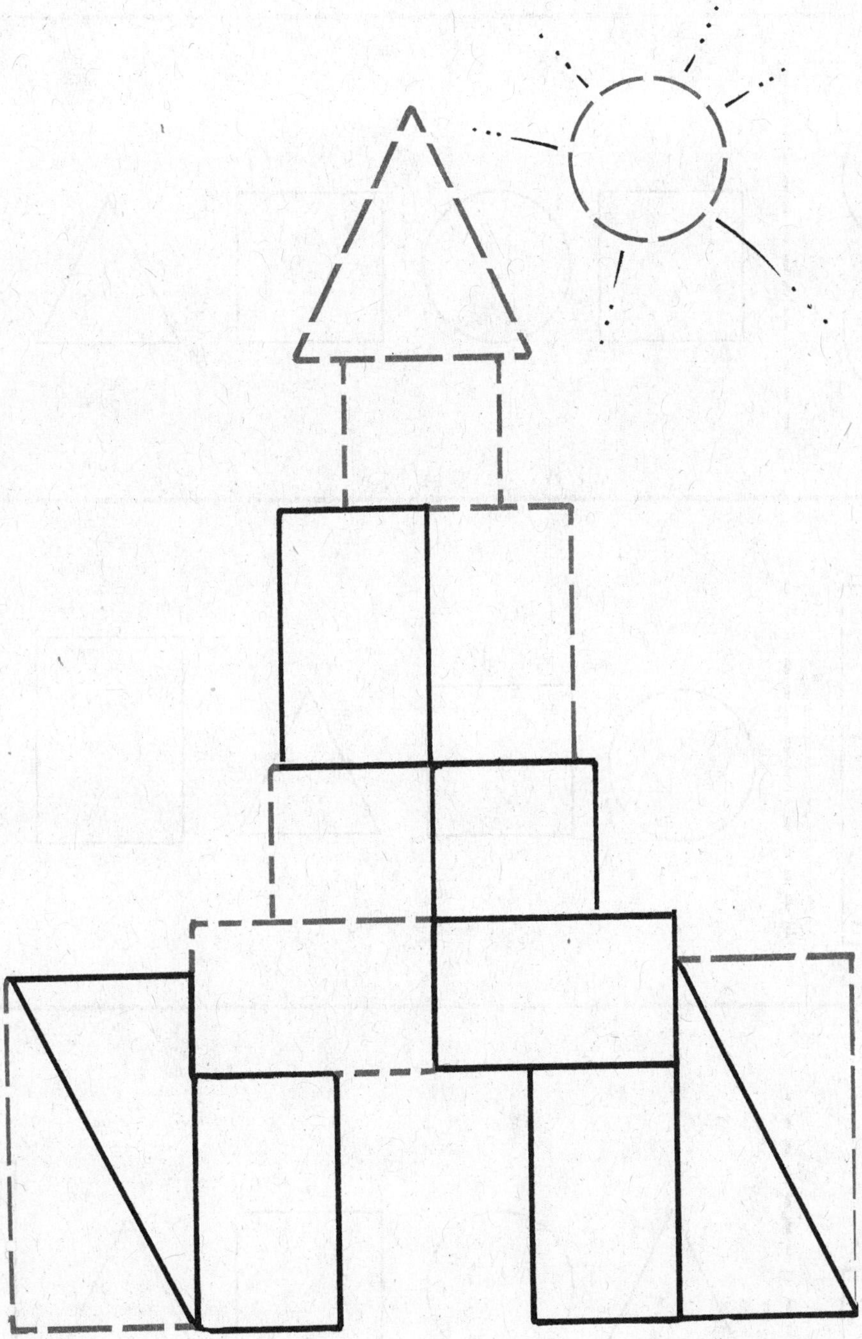

Use plane shapes to finish the puzzle. Use the same color to color the shapes that are alike.

Combine Plane Shapes

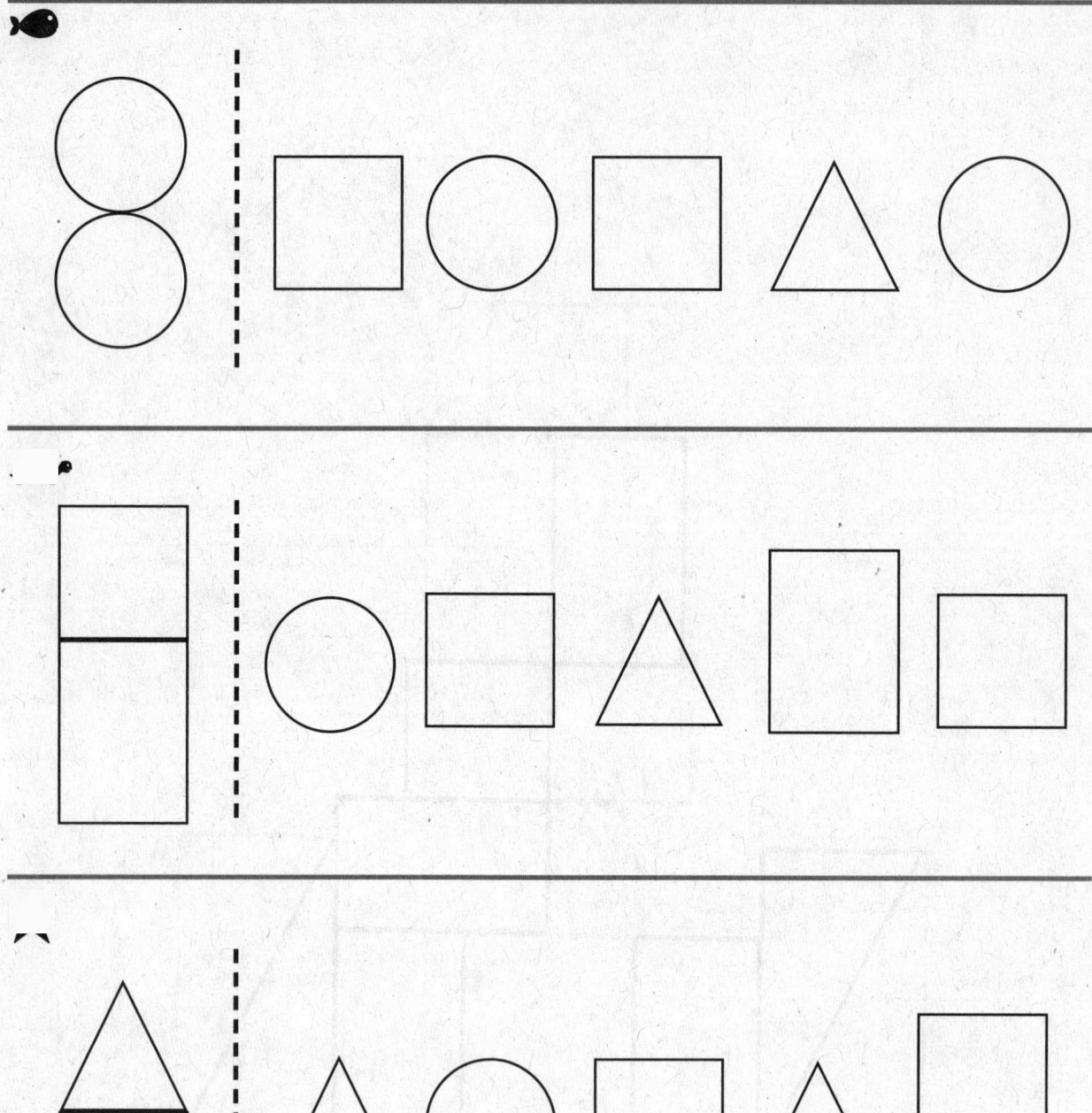

★ Use plane shapes to make each figure. Color the plane shapes you used.

Represent Plane Shapes

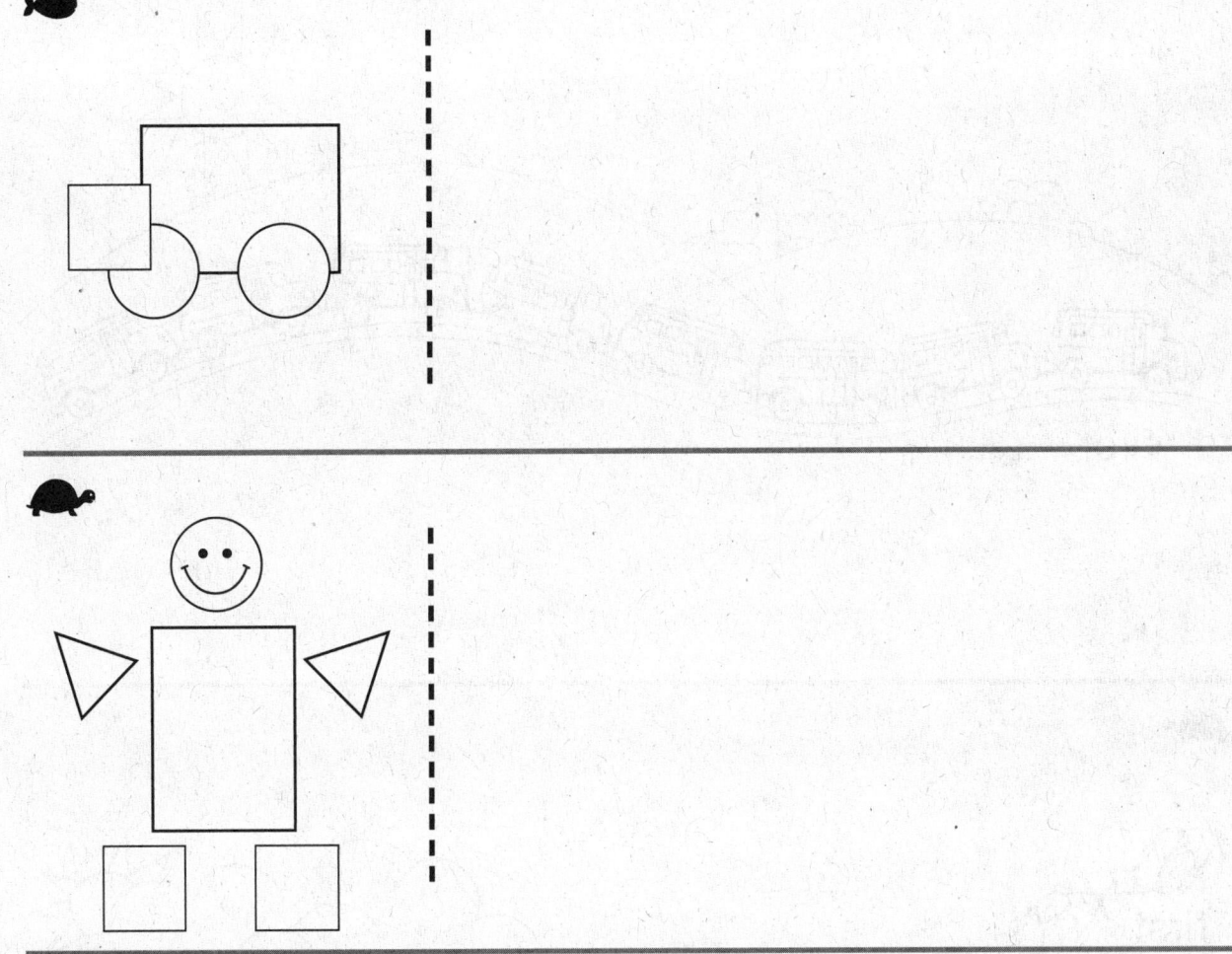

Look at the object at the beginning of the row. Draw the shapes that make up the object.

Ordinal Numbers

first

first

 Circle the eighth bus. Draw a line under the third bus. Mark an X on the first bus.

 Circle the fifth airplane. Draw a line under the tenth airplane. Mark an X on the fourth airplane.

Problem Solving Strategy: Use Objects

 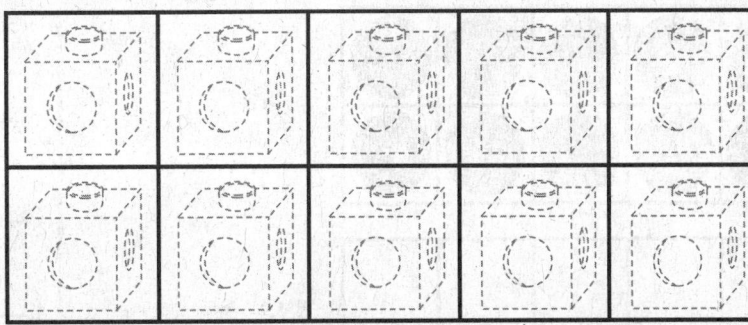 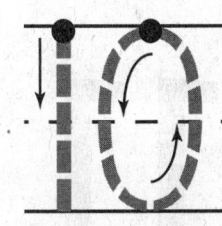

 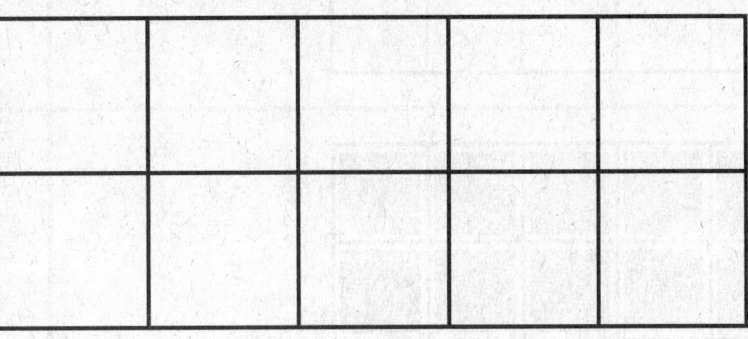 

 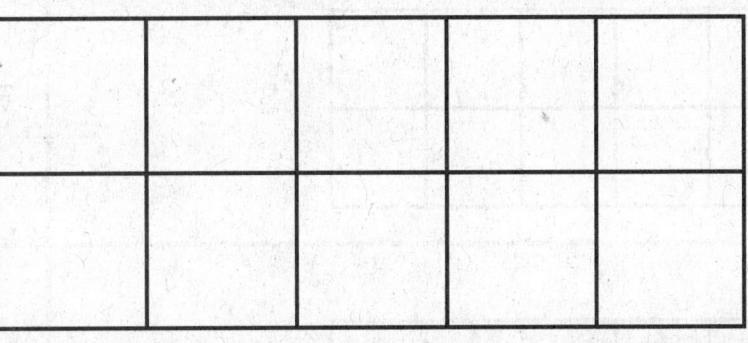 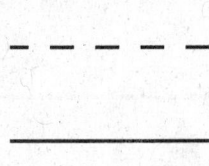

 Place ten cubes on the ten frame. Trace the number.

 Use cubes to model the number that is one more than nine. Write the number.

★ Use cubes to model the number that is one less than ten. Write the number.

♥ Use cubes to model the number that is two more than eight. Write the number.

11, 12, 13

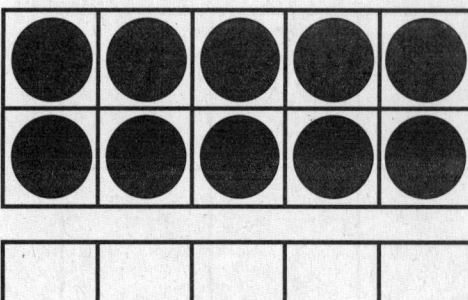

11

12

13

—————— · —————— — — — — — · —————— ——————

_____ 🐟 🐢 ★ Draw more shapes to show the number. Write the number.

Name _____

14, 15, 16

🐟

14

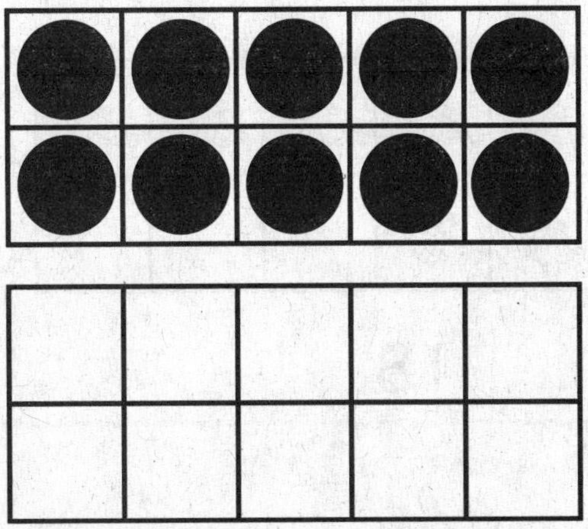

15

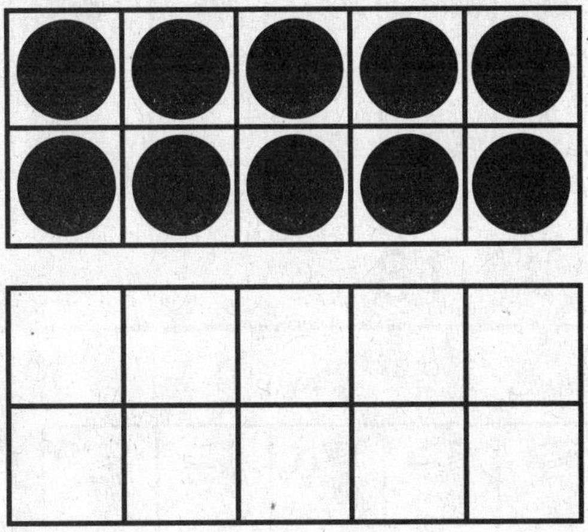

★

16

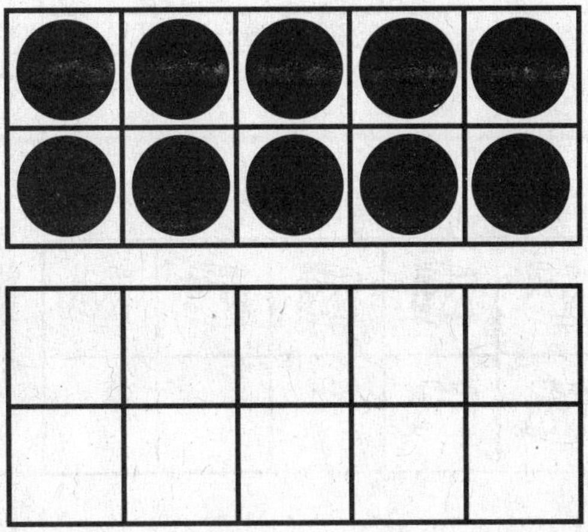

🐟 🐢 ★ Draw more counters to show the number. Write the number.

Name _____

17 and 18

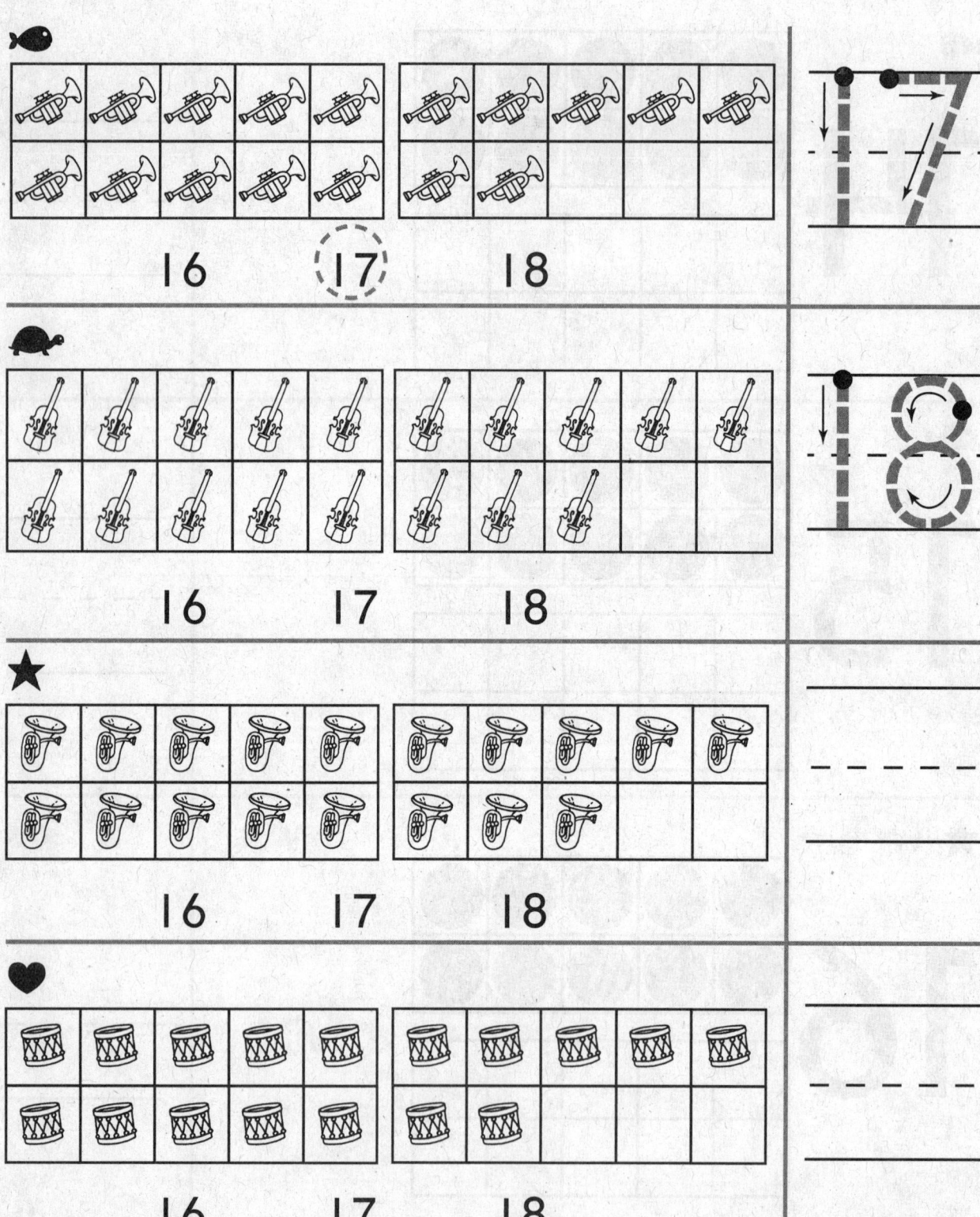

16   (17)   18

16   17   18

16   17   18

16   17   18

🐟 🐢 Count the instruments. Circle the number that tells how many. Trace the number.

★ ♥ Count the instruments. Circle the number that tells how many. Write the number.

Problem Solving Skill: Use Data from a Graph

Toys

Shapes

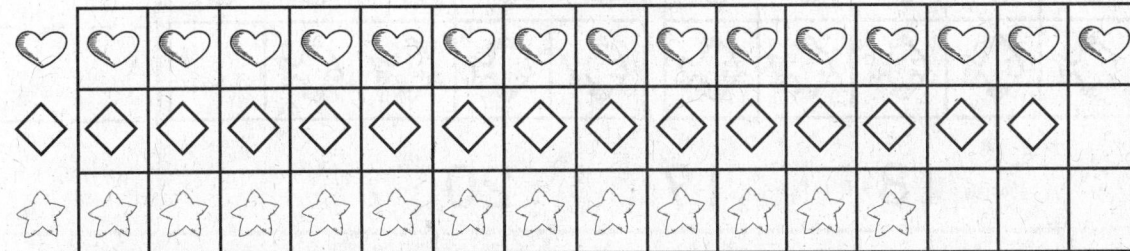

Count and write how many. Circle the number that shows the most.
Mark an X on the number that shows the fewest.

19 and 20

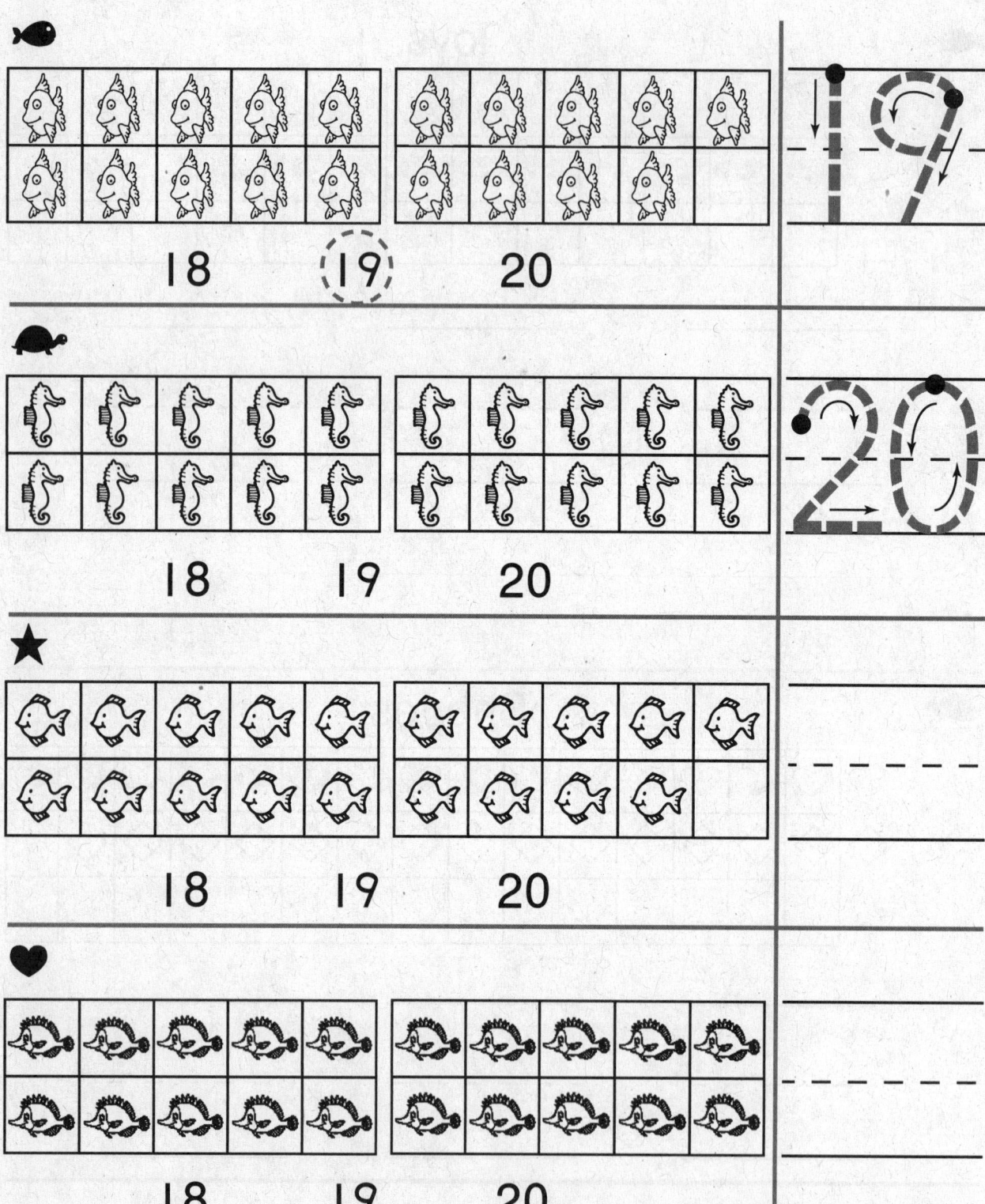

18   (19)   20

18   19   20

18   19   20

18   19   20

🐟 🐢 Count the fish. Circle the number that tells how many. Trace the number.
★ ♥ Count the fish. Circle the number that tells how many. Write the number.

Name _____

21 to 30

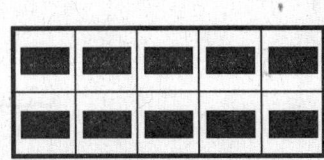 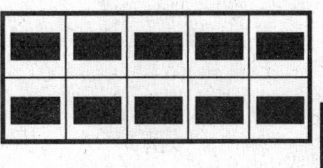 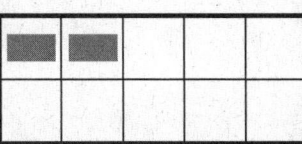

22

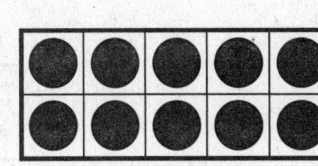 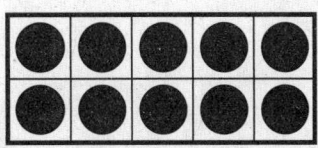 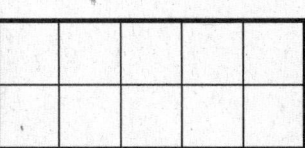

24

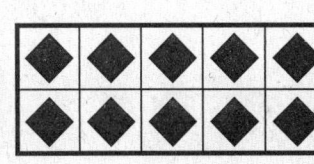 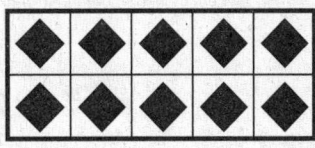 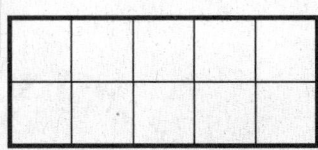

26

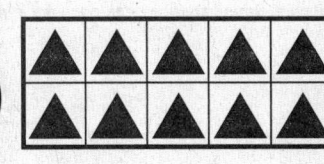 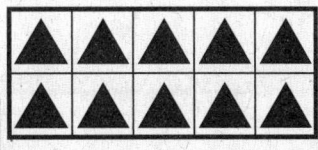 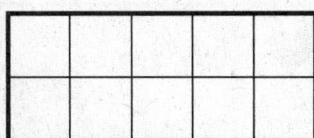

28

 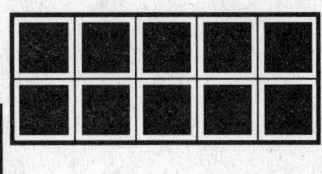

30

Draw more shapes to show the number.

© Harcourt

Penny

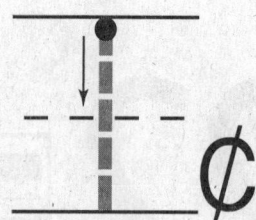

_____ ¢

_____ ¢

_____ ¢

_____ ¢

🐟 🐢 ★ ♥ Count the pennies. Write how many cents.

© Harcourt

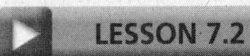

Nickel

   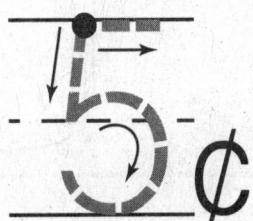

   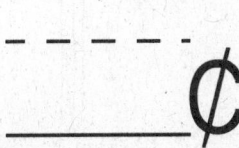

      Write how many cents.

Name _____

Dime

 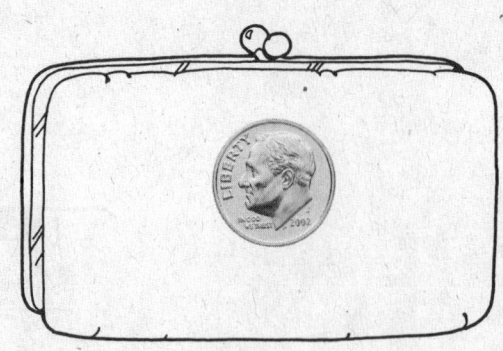

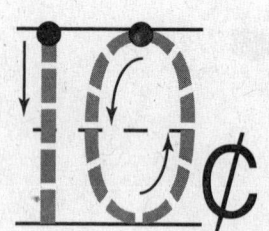

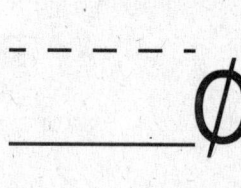

 🐋 ★ ♥ Write how many cents.

Quarter

5¢   10¢   25¢

5¢   10¢   25¢

★

5¢   10¢   25¢

♥

5¢   10¢   25¢

🐟 🐢 ★ ♥   Circle how many cents.

© Harcourt

Trade Money

  10¢

 5¢

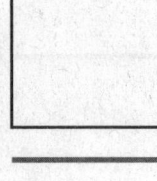  25¢

🐟 Trade a dime for the pennies. Draw the dime.

🐢 Trade pennies for the nickel. Draw the pennies.

★ Trade pennies for the quarter. Draw the pennies.

Name _____

Problem Solving Strategy: Use Objects

10¢ 8¢ 7¢ 5¢

🐟

_____

- - - - - - - -

_____ ¢

🐢

_____

- - - - - - - -

_____ ¢

🐟 Choose a group of balloons to buy. Circle the group. Draw the
pennies you would use to buy each group. Write how many cents.

🐢 Trade your pennies. Use other coins to show the same amount.

Name _____

Morning, Afternoon, Evening

Morning

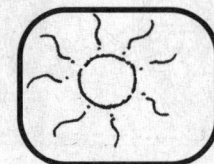

Afternoon

Evening

Morning

Afternoon

Evening

 What times of the day do the pictures show? Circle the time of day that is missing.

First, Second, Third

_____   _____   _____

- - - - - - -   - - - - - - -   - - - - - - -

_____   _____   _____

- - - - - - -   - - - - - - -   - - - - - - -

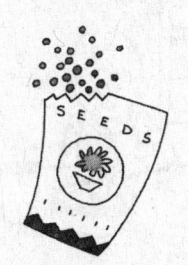  

_____   _____   _____

- - - - - - -   - - - - - - -   - - - - - - -

   Write 1, 2, or 3 to show what happens first, second and third.

Name _____

Time to the Hour

_____   _____   _____

_ _ _ _   _ _ _ _   _ _ _ _

_____ o'clock   _____ o'clock   _____ o'clock

_____   _____   _____

_ _ _ _   _ _ _ _   _ _ _ _

_____ o'clock   _____ o'clock   _____ o'clock

  Write the number that tells the hour.

★ Circle the time that each event happens.

© Harcourt

Days of the Week

February

Sunday	Monday	Tuesday	Wednesday	Thursday	Friday	Saturday
					1	2
3	4	5	6	7	8	9
10	11	12	13	14	15	16
17	18	19	20	21	22	23
24	25	26	27	28	29	

Monday   Wednesday   Friday

_____

Tuesday   Thursday   Friday

_____

★

Monday   Tuesday   Friday

_____

 Use yellow to color the days that hamburgers are served and circle the day of the week.

 Use red to color the days that spaghetti is served and circle the day of the week.

★ Use green to color the days that soup is served and circle the day of the week.

Months of the Year

January	1
Sun	Mon	Tue	Wed	Thu	Fri	Sat
	1	2	3	4	5
6	7	8	9	10	11	12
13	14	15	16	17	18	19
20	21	22	23	24	25	26
27	28	29	30	31

February	2
Sun	Mon	Tue	Wed	Thu	Fri	Sat
					1	2
3	4	5	6	7	8	9
10	11	12	13	14	15	16
17	18	19	20	21	22	23
24	25	26	27	28	29

March	3
Sun	Mon	Tue	Wed	Thu	Fri	Sat
						1
2	3	4	5	6	7	8
9	10	11	12	13	14	15
16	17	18	19	20	21	22
23/30	24/31	25	26	27	28	29

April	4
Sun	Mon	Tue	Wed	Thu	Fri	Sat
	1	2	3	4	5
6	7	8	9	10	11	12
13	14	15	16	17	18	19
20	21	22	23	24	25	26
27	28	29	30

May	5
Sun	Mon	Tue	Wed	Thu	Fri	Sat
				1	2	3
4	5	6	7	8	9	10
11	12	13	14	15	16	17
18	19	20	21	22	23	24
25	26	27	28	29	30	31

June	6
Sun	Mon	Tue	Wed	Thu	Fri	Sat
1	2	3	4	5	6	7
8	9	10	11	12	13	14
15	16	17	18	19	20	21
22	23	24	25	26	27	28
29	30

July	7
Sun	Mon	Tue	Wed	Thu	Fri	Sat
		1	2	3	4	5
6	7	8	9	10	11	12
13	14	15	16	17	18	19
20	21	22	23	24	25	26
27	28	29	30	31

August	8
Sun	Mon	Tue	Wed	Thu	Fri	Sat
					1	2
3	4	5	6	7	8	9
10	11	12	13	14	15	16
17	18	19	20	21	22	23
24/31	25	26	27	28	29	30

September	9
Sun	Mon	Tue	Wed	Thu	Fri	Sat
	1	2	3	4	5	6
7	8	9	10	11	12	13
14	15	16	17	18	19	20
21	22	23	24	25	26	27
28	29	30

October	10
Sun	Mon	Tue	Wed	Thu	Fri	Sat
			1	2	3	4
5	6	7	8	9	10	11
12	13	14	15	16	17	18
19	20	21	22	23	24	25
26	27	28	29	30	31

November	11
Sun	Mon	Tue	Wed	Thu	Fri	Sat
						1
2	3	4	5	6	7	8
9	10	11	12	13	14	15
16	17	18	19	20	21	22
23/30	24	25	26	27	28	29

December	12
Sun	Mon	Tue	Wed	Thu	Fri	Sat
	1	2	3	4	5	6
7	8	9	10	11	12	13
14	15	16	17	18	19	20
21	22	23	24	25	26	27
28	29	30	31

         ♥

_____   _____   _____   _____

- - - - -   - - - - -   - - - - -   - - - - -

_____   _____   _____   _____

🐟 What month is August? Write the number.

🐢 What month comes just before May? Say the name. Write the number.

★ What is the last month of the year? Say the name. Write the number.

♥ What month comes just after June? Say the name. Write the number.

Problem Solving Skill: Use a Calendar

October

Sunday	Monday	Tuesday	Wednesday	Thursday	Friday	Saturday
					1	2
3	4	5	6	7	8	9
10	11	12	13	14	15	16
17	18	19	20	21	22	23
24	25	26	27	28	29	30
31						

_____ Tuesdays

_____ Sundays

_____ Fridays

_____ Days in October

- Count the Tuesdays. Write how many.
- Count the Sundays. Write how many.
- ★ Count the Fridays. Write how many.
- ♥ Write how many days in October.

© Harcourt

Name _____

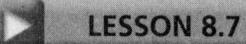

Seasons of the Year

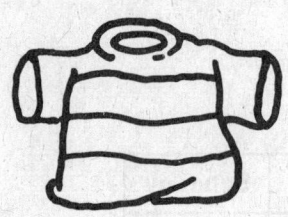   

spring **summer** **fall** **winter**

    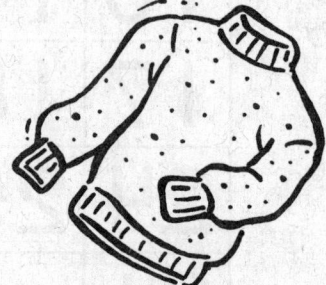

 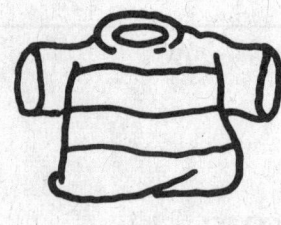   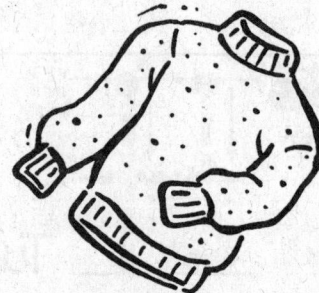

 Circle the shirt for fall.

 Circle the shirt for summer.

Name _____

Compare Lengths

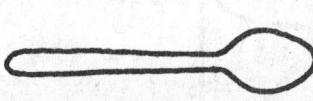

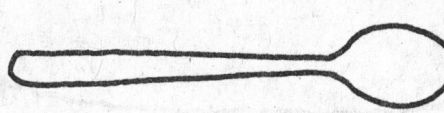

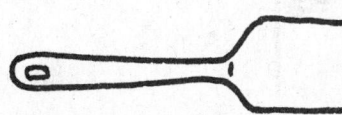

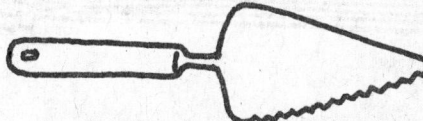

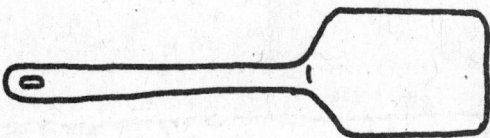

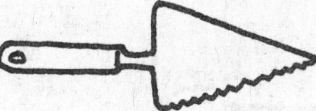

© Harcourt

🐟 🐢 ★ ♥ Circle the longer object. Draw a line under the shorter object.

Order Lengths

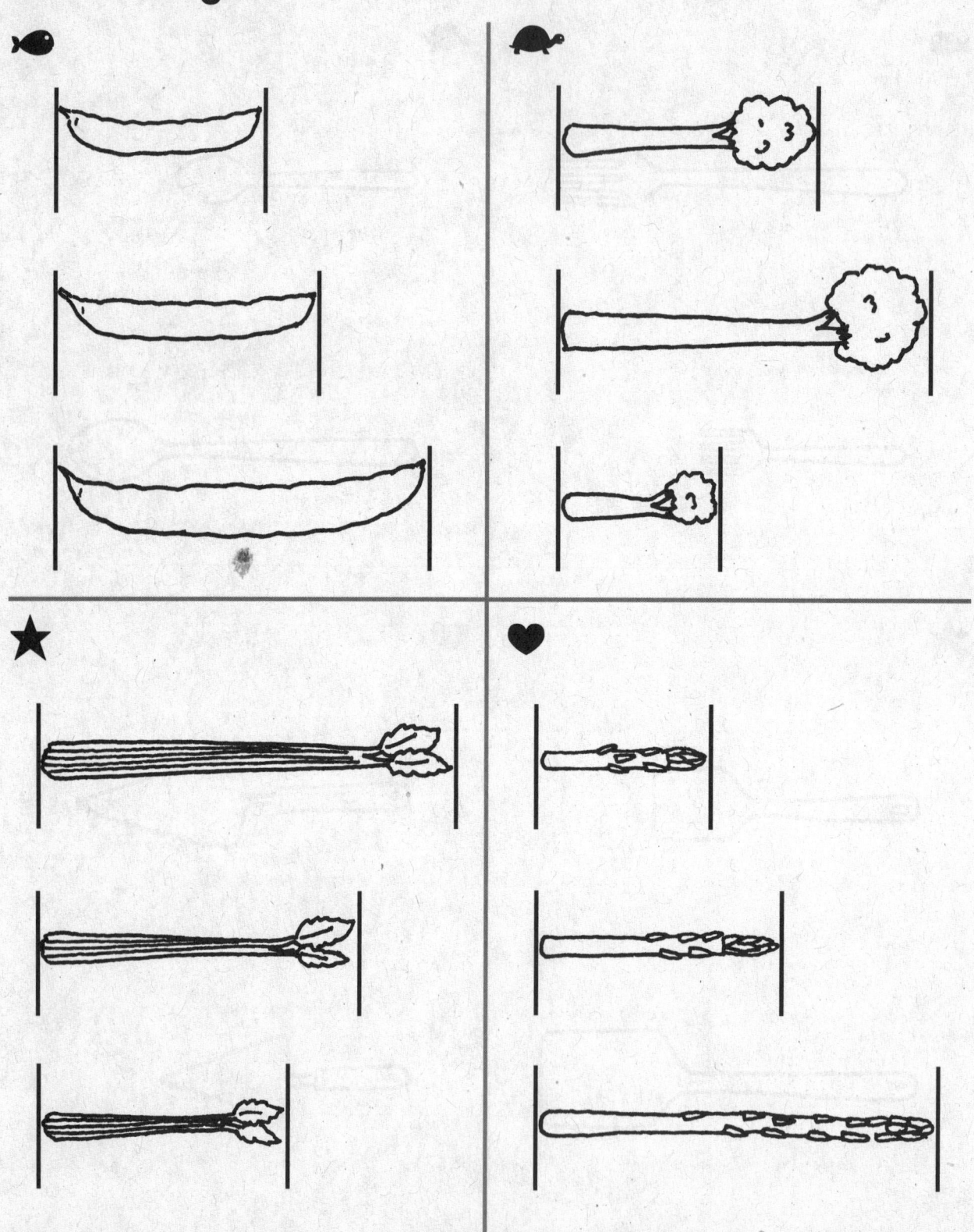

🐟 🐢 ★ ♥ Circle the groups of objects that are in order from shortest to longest, starting at the top.

PW 62 **Practice/Homework**

Name _____

Compare Height

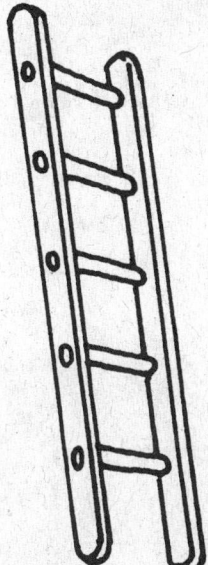

  ★ Circle the object that is shorter.
Draw a line under the object that is taller.

© Harcourt

Problem Solving Strategy: Use Objects

_ _ _ _ _ _   _ _ _ _ _ _   _ _ _ _ _ _

_____   _____   _____

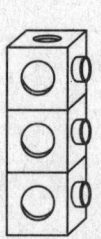 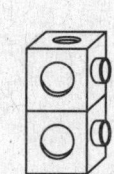 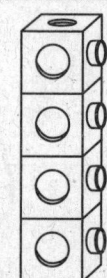

_ _ _ _ _ _   _ _ _ _ _ _   _ _ _ _ _ _

_____   _____   _____

★

_ _ _ _ _ _   _ _ _ _ _ _   _ _ _ _ _ _

_____   _____   _____

🐟 🐢 ★ Put the objects in order from shortest to tallest. Write 1, 2, and 3 to show the order. Circle the group that is in order from shortest to tallest.

Compare and Order Capacity

_____   _____   _____

- - - - - - - - -   - - - - - - - - -   - - - - - - - - -

_____   _____   _____

🐟🐢 Circle the container that holds more. Mark an X on the container that holds less.

★ Write 1, 2, and 3 to show the order of cups from the one that holds the least to the one that holds the most.

Name _____

Compare and Order Weight

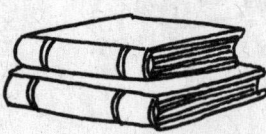  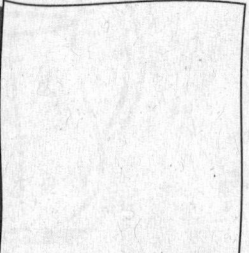

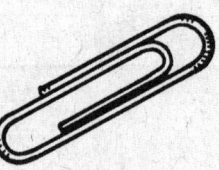  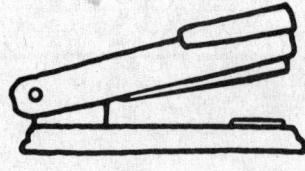

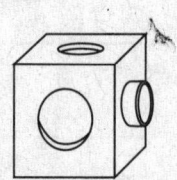  

★ Hold each object. Mark an X on the picture of the object that feels lightest. Underline the object that feels heaviest. Then circle the group of objects if they are in order from lightest to heaviest.

Concrete Graphs

🐟

How Many Blocks?

★

□ – – – – – – – – △ – – – – – – –
_____ _____

🐟 Place a handful of pattern blocks on the workspace. What can you ask a classmate about the blocks?

🐢 Make a graph with your pattern blocks.

★ Write how many of each pattern block. Circle the number that shows more blocks. Mark an X on the number that shows fewer blocks.

More Concrete Graphs

🐟 Children on the Playground

     _____

- 🐟 Look at the picture. What can you ask classmates about the boys and girls?
  Use red and blue cubes to make a graph. Draw each cube.
- 🐋 Write how many. Circle the number that is more.

PW68 Practice/Homework

Use Concrete Graphs

 ## Favorite Snack

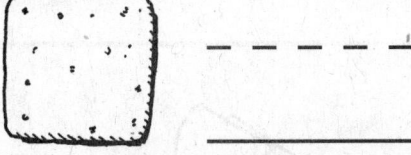  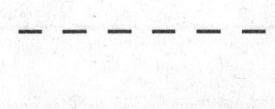

★

 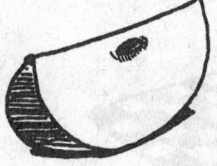

- Look at the snacks. What could you ask a classmate about the snacks? Ask five classmates. Place a counter to show each answer. Draw each counter.
- Write how many.
- ★ Circle the snack that shows fewer.

Name _____

Picture Graphs

 ## Indoor Activities

_____   _____   _____

   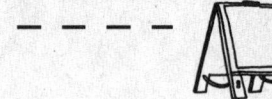   

_____   _____   _____

★

 What can you ask classmates about these activities? Ask five classmates. Draw a picture on the graph for each answer. Write how many.

🐢 Circle the picture that shows the activity the most children do.

★ Circle the picture that shows the activity the fewest children do.

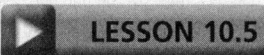

More Picture Graphs

On The Farm

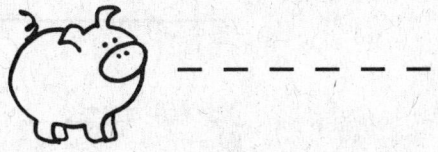   

 What can you ask classmates about the animals in the picture?
Make a picture graph.

 Write how many. Circle the number that shows fewer.

Name _____

Problem Solving Strategy: Use Data from a Graph

Summer Fun

   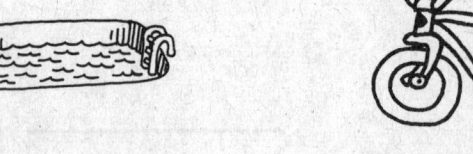   

_____   _____   _____

- - - - -   - - - - -   - - - - -

_____   _____   _____

- What can you ask classmates about these summer activities? Ask five classmates. Make a concrete graph or a picture graph.
- Write how many of each. Circle the one that shows the most.

Building 4 and 5

🐟 🐢 Use 2 different colors of bear counters to make the number 4. Write how many of each color you used to make 4.

★ ♥ Color to show a way to make 5. Write the numbers you used to make 5.

Building 6 and 7

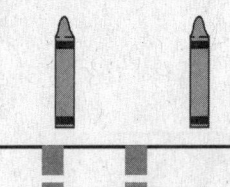  and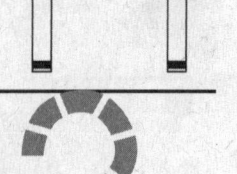

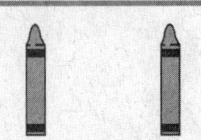  and

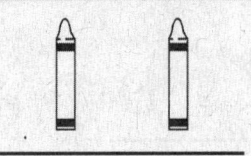  and

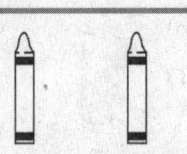  and

🐟 🐢 Count the gray and white crayons. Write the numbers used to make 6.
★ ♥ Color to show a way to make 7. Write the numbers you used to make 7.

Name _____

Building 8 and 9

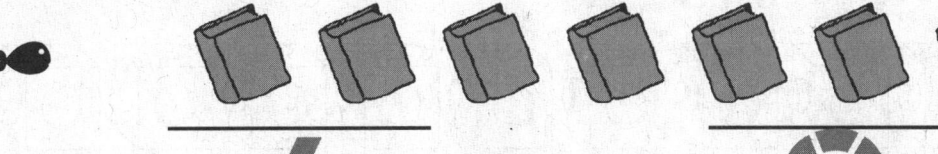

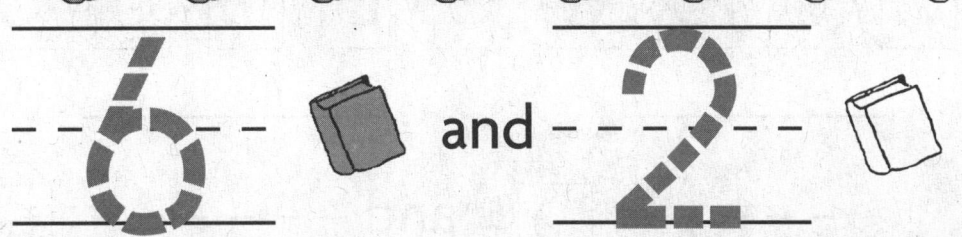

6 and 2

and

🐟 🐢 Count the gray and white books. Write the numbers used to build 8.
★ ♥ Color to show a way to make 9. Write the numbers you used to make 9.

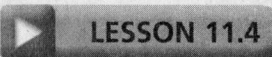
Name _____

Building 10

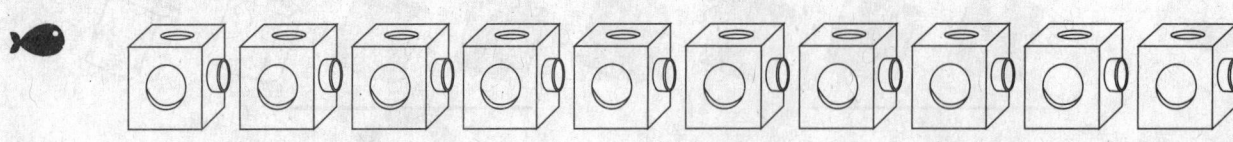

_____   _____

and

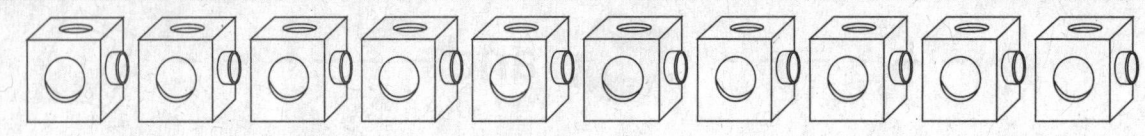

_____   _____

and

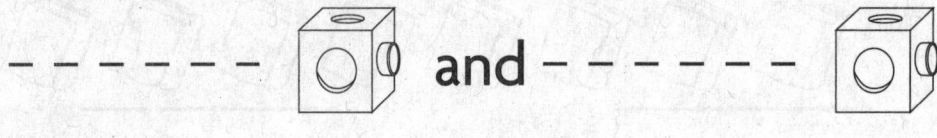

_____   _____

and

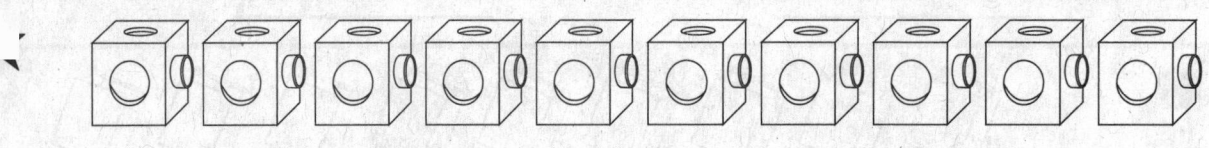

★ Color the cubes two different colors to show a way to make 10. Write the numbers you used to make 10.

© Harcourt

Problem Solving Strategy: Act It Out

_____

- - - - - - - - - - - -

_____

_____

- - - - - - - - - - - -

_____

 Tell an addition story about the picture. Act it out.
Write the number that tells how many children there are in all.

Model Addition

3 and 1 is 4

3 and 2 is _____

★

2 and 2 is _____

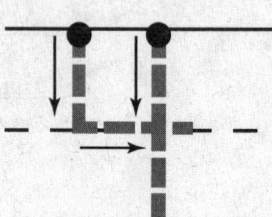 ★ Tell an addition story about the picture. Model the story with connecting cubes. Write the number that tells how many in all.

Name _____

Use Pictures to Add

3 and 1 is 4

_____   and   _____   is   _____

★

_____   and   _____   is   _____

  ★ Tell a story about the picture. Complete the addition sentence.

Name _____

Addition Stories

_ _ _ _ _ + _ _ _ _ _ = _ _ _ _ _

_ _ _ _ _ + _ _ _ _ _ = _ _ _ _ _

Tell an addition story. Model your story with objects.
Draw the objects and complete the addition sentence.

Name _____

Take Away

[blank box]

5 take away 1

_____
- - - - - - -
_____

[blank box] 🐢

6 take away 4

_____
- - - - - - -
_____

★ [blank box]

7 take away 4

_____
- - - - - - -
_____

🐟 🐢 ★ Tell a story. Model the story with connecting cubes.
Write the numbers that tells how many are left.

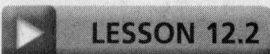

Problem Solving Strategy • Act It Out

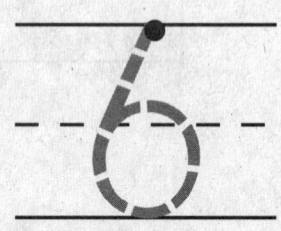

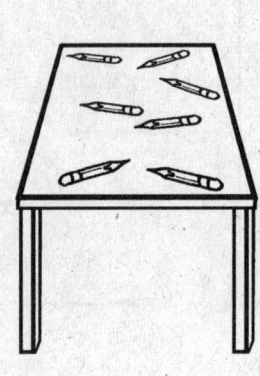

Tell a subtraction story about the picture. Act it out. Count the objects that are left. Write the number that tells how many objects are left.

© Harcourt

Model Subtraction

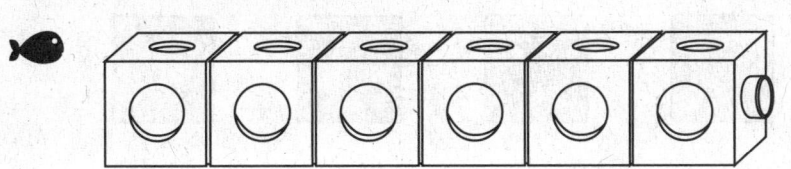

7 take away **1** is

_ _ _ _ _ _

_____

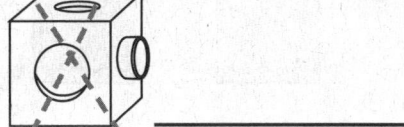

5 take away **1** is

_ _ _ _ _ _

_____

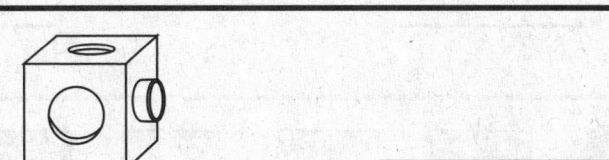

6 take away **1** is

_ _ _ _ _ _

_____

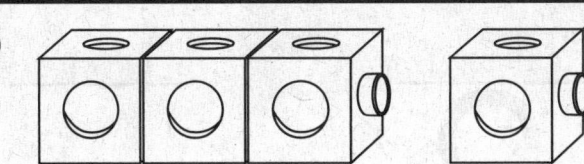

4 take away **1** is

_ _ _ _ _ _

_____

🐟 🐢 ★ ♥ Count the connecting cubes. Mark an X on the cube that is taken away.
Write the number that tells how many cubes are left.

Name _____

Use Objects to Subtract

  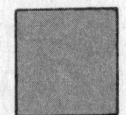 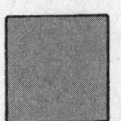   

_____   _____   _____

_____ take away _____ **2** is _____

_____   _____   _____

   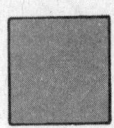  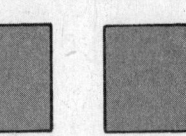   

_____   _____   _____

_____ take away _____ **4** is _____

_____   _____   _____

 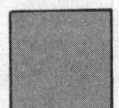 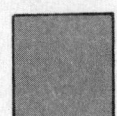      

_____   _____   _____

_____ take away _____ **2** is _____

_____   _____   _____

 Use color tiles to show how many there are in all. Write the number.
Mark an X as you take away color tiles. Write how many are left.

Use Pictures to Subtract

$$8 - 4 = \underline{\qquad}$$

$$9 - 3 = \underline{\qquad}$$

Tell the subtraction story. Then complete the subtraction sentence.

Subtract with Pennies

5¢ take away **2**¢ is **3**¢

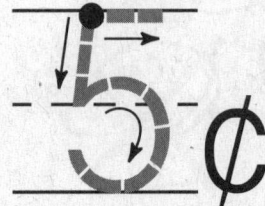       

_____¢ take away _____¢ is _____¢

_____¢ take away _____¢ is _____¢

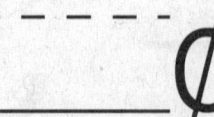 Tell a subtraction story. Complete the subtraction sentence to tell how many pennies are left.

Subtraction Stories

_____   _____   _____

‑ ‑ ‑ ‑ ‑ ‑ ‑ take away ‑ ‑ ‑ ‑ ‑ ‑ is ‑ ‑ ‑ ‑ ‑

_____   _____   _____

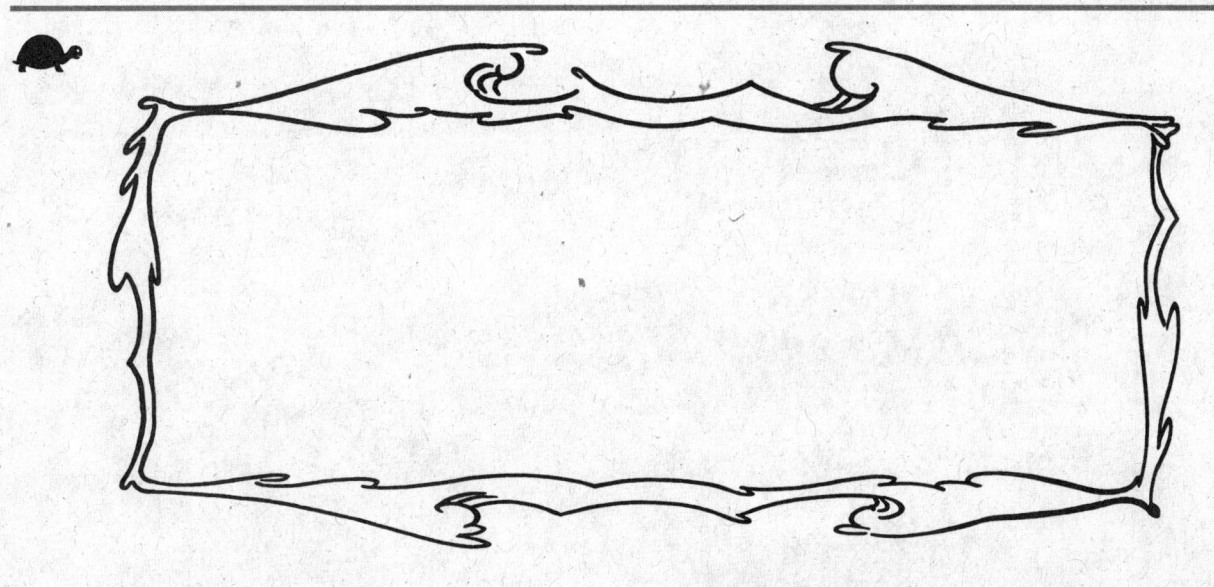

_____   _____   _____

‑ ‑ ‑ ‑ ‑ ‑ ‑ take away ‑ ‑ ‑ ‑ ‑ ‑ is ‑ ‑ ‑ ‑ ‑

_____   _____   _____

🐟 🐢 Tell a subtraction story. Act out your story with objects. Draw the objects, and mark an X on the objects you subtract. Complete the subtraction sentence.